TWO TRUTHS & ONE LIE!

MATH EDITION

101 DAILY ACTIVITIES FOR GRADES 3, 4, 5

mashupmath

Hey There!

Welcome to *Two Truths & One Lie: Math Edition* for grades 3, 4, and 5!

You're probably here because you're looking for fun ways to engage your kids at the start or end of every class— because you know that learning math was meant to be a fun, visual and thought-provoking experience!

Starting classes with *Two Truths and One Lie* (*2T1L*) activities is a great way to spark creative and critical student thinking that will last for an entire lesson and beyond.

2T1L activities help your kids to develop reasoning skills, make logical arguments, express their ideas in words, and engage with visual mathematics—which ultimately leads to deeper and more meaningful understanding of challenging topics and concepts.

The daily graphics found in this book can be applied to dozens of topics and are aligned with math learning standards typically covered in grades 3, 4, and 5.

 Enjoy!

How to Use This Book

Finding fun and engaging warm-up and/or exit activities for your kids doesn't have to be a struggle.

Using 2T1L activities is an effective way to boost student engagement and critical thinking. These activities work best when students can work collaboratively and have opportunities to justify their thinking via small group or whole class discussions.

The best way to use this book is to choose a graphic that aligns with topic(s) from the day's lesson and project it as large as you can at the front of your classroom. Then, have students:

- **Observe and think**
- **Choose which statement is a lie**
- **Write an explanation/justification**
- **Share with their classmates**

Be sure to make your students justify WHY their choice is a lie and why the others must be true!

Example (4th Grade)

Imagine your kids entering class expecting a typical warm-up exercise, only to see the 2T1L graphic below posted on the board.

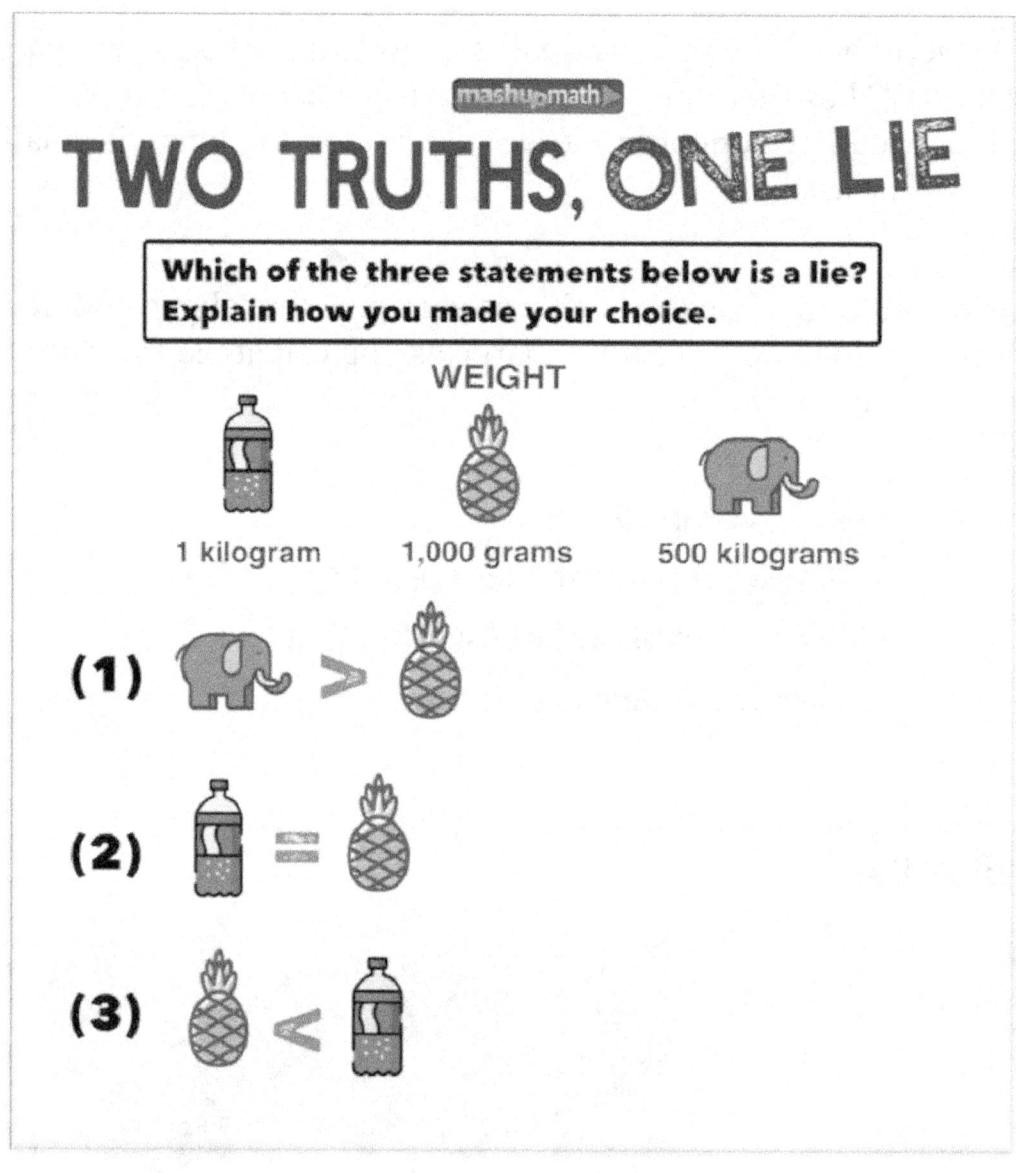

Sample Discussion

Student: I think that statement 3 is a lie because I know that statements 1 and 2 are true.

Teacher: How do you know that statements 1 and 2 are true?

Student: I know that 2 is true because one kilogram is equal to one thousand grams. And if statement 2 is true, then 1 must also be true because 500 kilograms equals 500,000 grams, which is heavier than 1,000 grams.

Teacher: And how can you be sure that statement 3 is a lie?

Student: If statement 2 is true and the soda bottle weighs the same as the pineapple, then saying that the pineapple weighs less is a lie!

Teacher: I agree with your justification! How could you change this problem to make statement 3 true and statement 1 or 2 false?

Standards Breakdown

3rd Grade
4th Grade
5th Grade

3rd Grade (Page 11)

Topic/Standard	Worksheets
Represent and solve problems involving multiplication and division. Interpret products of whole numbers, e.g., interpret 5 × 7 as the total number of objects in 5 groups of 7 objects each. Understand properties of multiplication and the relationship between multiplication and division. Determine the unknown whole number in a multiplication or division equation relating three whole numbers.	1, 2, 3, 4, 5, 6, 7, 8, 9, 10
Tell and write time to the nearest minute and measure time intervals in minutes.	11, 12
Measure and estimate liquid volumes and masses of objects using standard units of grams (g), kilograms (kg), and liters (l).[1] Add, subtract, multiply, or divide to solve one-step word problems involving masses or volumes that are given in the same units, e.g., by using drawings (such as a beaker with a measurement scale) to represent the problem.	13, 14
Geometric measurement: understand concepts of area and relate area to multiplication and to addition. Measure areas by counting unit squares (square cm, square m, square in, square ft, and improvised units).	15, 16, 17, 18, 19, 20, 21, 22, 23
Develop understanding of fractions as numbers. Understand a fraction as a number on the number line; represent fractions on a number line diagram. Explain equivalence of fractions in special cases, and compare fractions by reasoning about their size. Compare two fractions with the same numerator or the same denominator by reasoning about their size.	24, 25, 26, 27, 28, 29, 30, 31
Represent and interpret data. Draw a scaled picture graph and a scaled bar graph to represent a data set with several categories. Solve one- and two-step "how many more" and "how many less" problems using information presented in scaled bar graphs. *For example, draw a bar graph in which each square in the bar graph might represent 5 pets.* Generate measurement data by measuring lengths using rulers marked with halves and fourths of an inch. Show the data by making a line plot, where the horizontal scale is marked off in appropriate units— whole numbers, halves, or quarters.	32, 33, 34,

4th Grade (Page 47)	
Topic/Standard	**Worksheets**
Numbers and operations in base-10. Recognize that in a multi-digit whole number, a digit in one place represents ten times what it represents in the place to its right. *For example, recognize that 700 ÷ 70 = 10 by applying concepts of place value and division.*	1, 2, 3, 4, 5,
Solve problems involving measurement and conversion of measurements. Apply the area and perimeter formulas for rectangles in real world and mathematical problems. *For example, find the width of a rectangular room given the area of the flooring and the length, by viewing the area formula as a multiplication equation with an unknown factor.*	6,7,8,9,10
Multiply a whole number of up to four digits by a one-digit whole number, and multiply two two-digit numbers, using strategies based on place value and the properties of operations. Illustrate and explain the calculation by using equations, rectangular arrays, and/or area models. Find whole-number quotients and remainders with up to four-digit dividends and one-digit divisors, using strategies based on place value, the properties of operations, and/or the relationship between multiplication and division. Illustrate and explain the calculation by using equations, rectangular arrays, and/or area models.	11, 12, 13, 14, 15
Draw and identify lines and angles, and classify shapes by properties of their lines and angles.	16
Extend understanding of fraction equivalence and ordering. Compare two fractions with different numerators and different denominators. Build fractions from unit fractions. Understand decimal notation for fractions, and compare decimal fractions.	17, 18, 19, 20, 21, 22, 23, 24, 25, 26, 27, 28, 29, 30
Use the four operations to solve word problems involving distances, intervals of time, liquid volumes, masses of objects, and money, including problems involving simple fractions or decimals, and problems that require expressing measurements given in a larger unit in terms of a smaller unit. Represent measurement quantities using diagrams such as number line diagrams that feature a measurement scale.	31, 32, 33, 34, 35

5th Grade (Page 83)	
Topic/Standard	**Worksheets**
Understand the place value system. Explain patterns in the number of zeros of the product when multiplying a number by powers of 10, and explain patterns in the placement of the decimal point when a decimal is multiplied or divided by a power of 10. Use whole-number exponents to denote powers of 10. Compare two decimals to thousandths based on meanings of the digits in each place, using >, =, and < symbols to record the results of comparisons. Perform operations with multi-digit whole numbers and with decimals to hundredths.	1, 2, 3, 4, 5, 6, 7, 8
Write and interpret numerical expressions. Use parentheses, brackets, or braces in numerical expressions, and evaluate expressions with these symbols.	9, 10, 11
Find whole-number quotients of whole numbers with up to four-digit dividends and two-digit divisors, using strategies based on place value, the properties of operations, and/or the relationship between multiplication and division. Illustrate and explain the calculation by using equations, rectangular arrays, and/or area models.	12, 13, 14
Fluently multiply multi-digit whole numbers using the standard algorithm. Convert like measurement units within a given measurement system. **Represent and Interpret Data.**	16, 17, 18, 19
Use equivalent fractions as a strategy to add and subtract fractions. Add and subtract fractions with unlike denominators Apply and extend previous understandings of division to divide unit fractions by whole numbers and whole numbers by unit fractions.	20, 21, 22, 23, 24, 25, 26, 27
Recognize volume as an attribute of solid figures and understand concepts of volume measurement. Measure volumes by counting unit cubes, using cubic cm, cubic in, cubic ft, and improvised units.	28, 29, 30, 31, 32
Represent real world and mathematical problems by graphing points in the first quadrant of the coordinate plane, and interpret coordinate values of points in the context of the situation.	33, 34, 35

3RD GRADE

TWO TRUTHS, ONE LIE

Which of the three statements below is a lie? Explain how you made your choice.

(1) $10 \times 6 = 10 \times 3 + 5 \times 3$

(2) $8 \times 10 = 8 \times 5 + 8 \times 5$

(3) $9 \times 10 = 4 \times 10 + 5 \times 10$

TWO TRUTHS, ONE LIE

Which of the three statements below is a lie? Explain how you made your choice.

(1) $10 \times 7 = 10 \times 4 + 10 \times 3$

(2) $9 \times 5 = 8 \times 5 + 1 \times 5$

(3) $12 \times 10 = 8 \times 10 + 12 \times 4$

TWO TRUTHS, ONE LIE

Which of the three statements below is a lie? Explain how you made your choice.

(1) 7 × 2 = 7 × 1 + 7 × 1

(2) 9 × 9 = 7 × 7 + 2 × 2

(3) 10 × 6 = 3 × 6 + 7 × 6

TWO TRUTHS, ONE LIE

Which of the three statements below is a lie? Explain how you made your choice.

(1) (7 × 🐻) = (5 × 🐻) + (2 × 🐻)

(2) (🐟 × 10) = (🐟 × 5) + (🐟 × 5)

(3) (1 × 🐹) = (2 × 🐹) + (0 × 🐹)

Hint: Pick a non-zero number, any non-zero number…

TWO TRUTHS, ONE LIE

Which of the three statements below is a lie? Explain how you made your choice.

(1) $4 + 4 + 4 + 4$

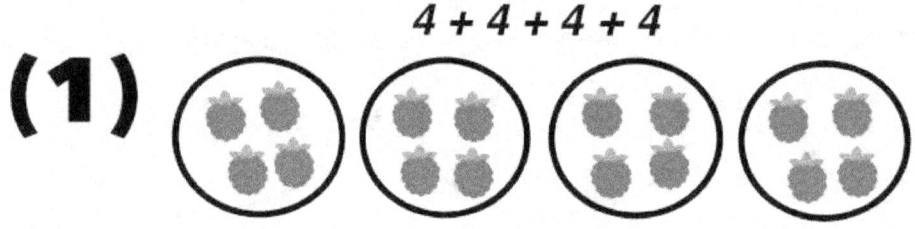

(2) 3 times 2

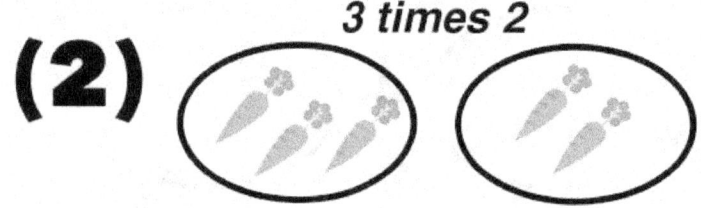

(3) 5 groups of three

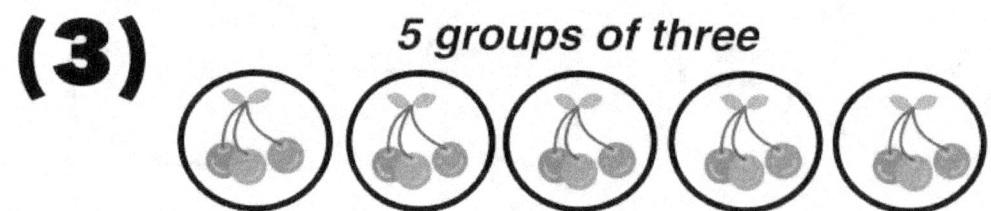

TWO TRUTHS, ONE LIE

Which of the three statements below is a lie? Explain how you made your choice.

(1) 5 groups of four

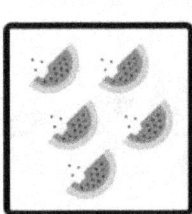  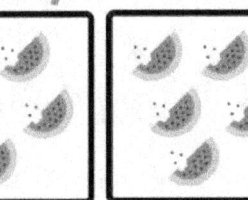 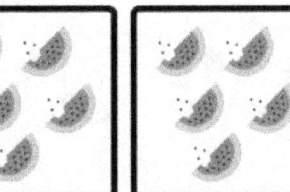

(2)  × 3 =  +  +

(3) 3 groups of three

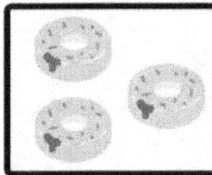 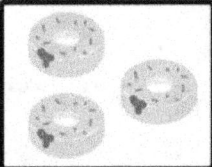 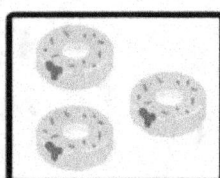

TWO TRUTHS, ONE LIE

Which of the three statements below is a lie? Explain how you made your choice.

(1) 🧑‍🚀 × 5 = 🧑‍🚀 + 🧑‍🚀 + 🧑‍🚀 + 🧑‍🚀 + 🧑‍🚀 + 🧑‍🚀

(2) 🚀 × 0 = 0

(3) 👽 + 👽 + 👽 + 👽 + 👽 + 👽 + 👽 = 7 × 👽

TWO TRUTHS, ONE LIE

Which of the three statements below is a lie? Explain how you made your choice.

(1) [array of dots in 3 rows of 4] = 12 ÷ 4

(2) [array of dots in 4 columns of 3] = 12 ÷ 4

(3) [array of dots divided into sections] = 12 ÷ 12

TWO TRUTHS, ONE LIE

Which of the three statements below is a lie? Explain how you made your choice.

(1) 4 times 4

(2) 4 times 1

 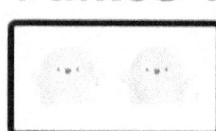 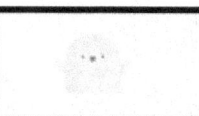

(3) 3 times 6

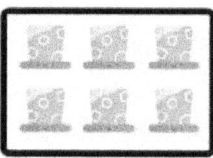  

TWO TRUTHS, ONE LIE

Which of the three statements below is a lie? Explain how you made your choice.

(1) ▶ $6 \div 2$

(2)  ▶ The cones are divided into five equal groups.

(3) 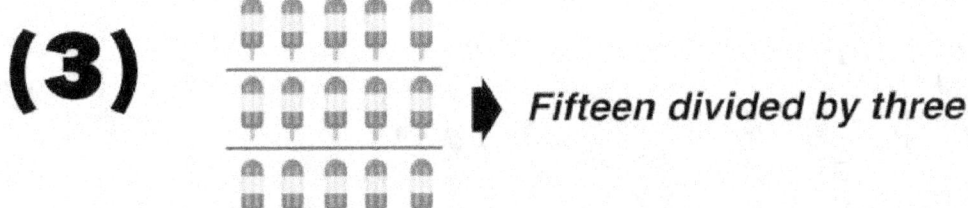 ▶ Fifteen divided by three

TWO TRUTHS, ONE LIE

Which of the three statements below is a lie? Explain how you made your choice.

(1)

(2)  4:39

(3)

TWO TRUTHS, ONE LIE

Which of the three statements below is a lie? Explain how you made your choice.

(1) 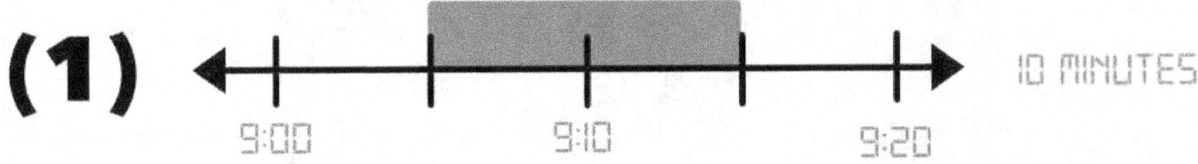 10 MINUTES

(2)  1 MINUTE

(3)  10 MINUTES

TWO TRUTHS, ONE LIE

Which of the three statements below is a lie? Explain how you made your choice.

300 GRAMS   1 KILOGRAM   400 GRAMS

(1)

(2)

(3)

TWO TRUTHS, ONE LIE

Which of the three statements below is a lie? Explain how you made your choice.

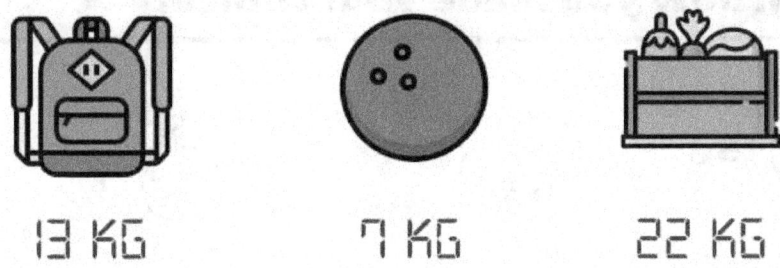

13 KG   7 KG   22 KG

(1) The vegetable crate is 9 kilograms heavier than the backpack.

(2) One crate of vegetables weighs less than three bowling balls.

(3) A bowling ball is 6 kilograms lighter than the backpack.

TWO TRUTHS, ONE LIE

Which of the three statements below is a lie? Explain how you made your choice.

(1)  Total = 16

(2)  Total = 24

(3)  Total = 49

TWO TRUTHS, ONE LIE

Which of the three statements below is a lie? Explain how you made your choice.

(1) 🐵 | 6 ⟶ 24   🐵 = 18

(2) 🐄 | 18 | 🐄 ⟶ 30   🐄 = 12

(3) 26 | 🐻 ⟶ 50   🐻 = 24

TWO TRUTHS, ONE LIE

Which of the three statements below is a lie? Explain how you made your choice.

15 — 🐤 — 🐼 — 60 — 🐮 — 90 — 🐱

(1) 🐤 = 30 & 🐱 = 100

(2) 🐼 = 45 & 🐮 = 75

(3) 🐮 = 75 & 🐱 = 105

TWO TRUTHS, ONE LIE

Which of the three statements below is a lie? Explain how you made your choice.

(1) $54 \div 6 = 30 \div 6 + 24 \div 6$

(2) $49 \div 7 = 35 \div 7 + 14 \div 7$

(3) $64 \div 8 = 42 \div 8 + 24 \div 8$

TWO TRUTHS, ONE LIE

Which of the three statements below is a lie? Explain how you made your choice.

🍉 = 3   💍 = 5   🐻 = 6

(1) 🍉 × 7 = 21

(2) 30 = 💍 × 6

(3) 7 × 🐻 = 49

TWO TRUTHS, ONE LIE

Which of the three statements below is a lie? Explain how you made your choice.

🐢 = 3   🐋 = 9   🦀 = 6

(1) $36 \div 🦀 = 6$

(2) $63 \div 🐋 = 8$

(3) $🐋 = 27 \div 🐢$

TWO TRUTHS, ONE LIE

Which of the three statements below is a lie? Explain how you made your choice.

Each ☐ is one square unit in all of the figures below.

(1) These figures have the same area.

(2) These figures have the same Area.

(3) 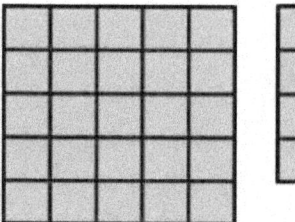 These figures have the same Area.

TWO TRUTHS, ONE LIE

Which of the three statements below is a lie? Explain how you made your choice.

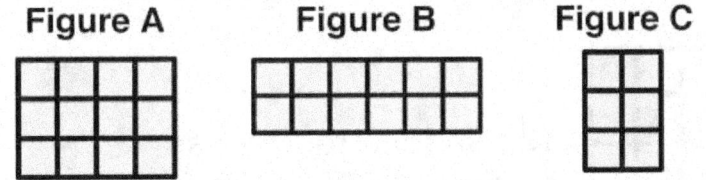

Figure A   Figure B   Figure C

(1) Figure A has a larger area than Figure B.

(2) Figure B has a larger area than Figure C.

(3) Figure C has an area of 6 square units.

TWO TRUTHS, ONE LIE

Which of the three statements below is a lie? Explain how you made your choice.

(1) 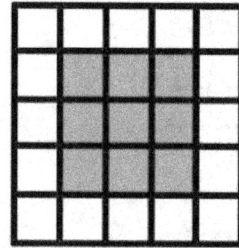 The shaded region has an area of 25 square units.

(2) 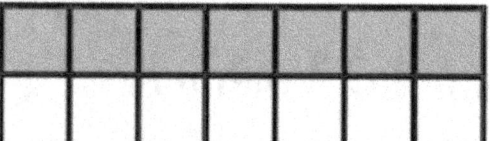 Both the shaded and the non-shaded regions have an area of 7 square units.

(3) 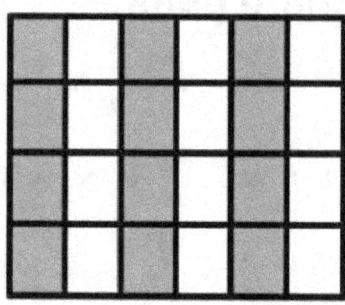 The total area of the figure is 24 square units.

TWO TRUTHS, ONE LIE

Which of the three statements below is a lie? Explain how you made your choice.

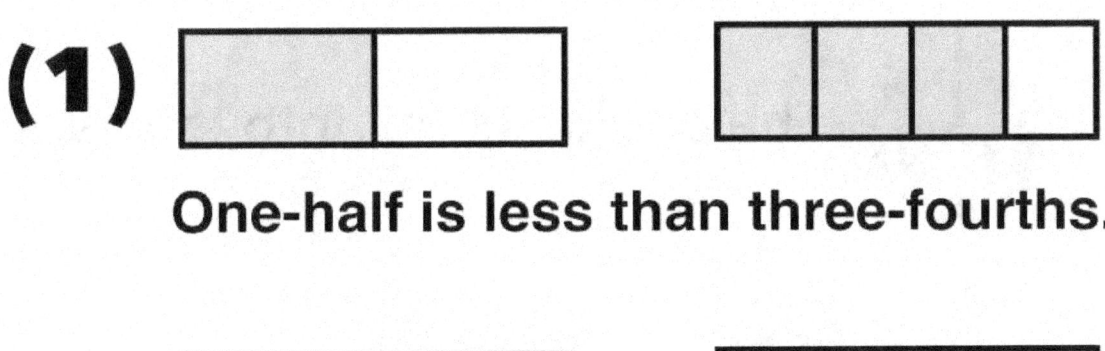

(1) One-half is less than three-fourths.

(2) One-half is more than two-thirds.

(3) Three-fourths is equal to six-eighths.

TWO TRUTHS, ONE LIE

Which of the three statements below is a lie? Explain how you made your choice.

(1) 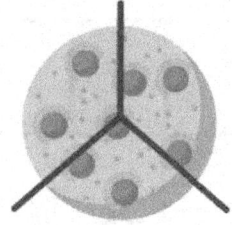 The cookie is cut into thirds.

(2)  The cake is cut in half.

(3) 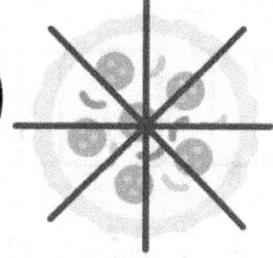 The pizza is cut into sixths.

TWO TRUTHS, ONE LIE

Which of the three statements below is a lie? Explain how you made your choice.

(1)  The image is cut into two equal parts.

(2) 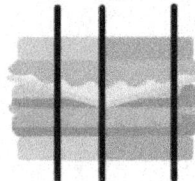 The image is cut into four equal parts.

(3) 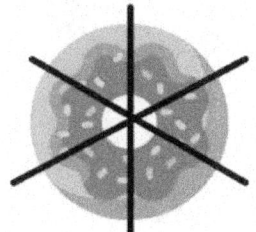 The image is cut into six equal parts.

TWO TRUTHS, ONE LIE

Which of the three statements below is a lie? Explain how you made your choice.

(1) 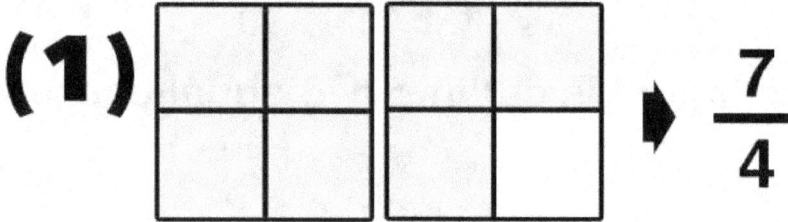 → $\dfrac{7}{4}$

(2) 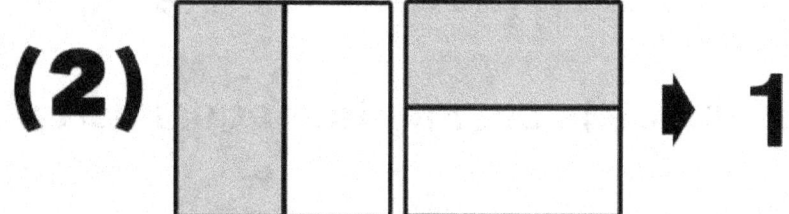 → 1

(3) 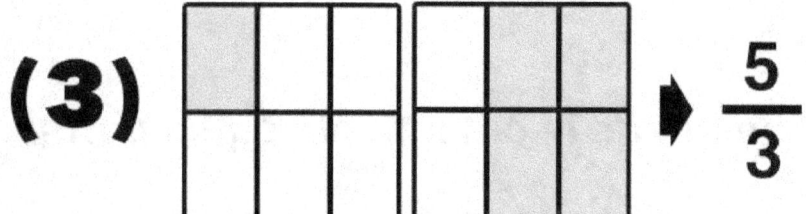 → $\dfrac{5}{3}$

TWO TRUTHS, ONE LIE

Which of the three statements below is a lie? Explain how you made your choice.

 $= \dfrac{1}{4}$  $= \dfrac{1}{8}$ (monster) $= \dfrac{1}{15}$

(1)  >

(2)  = one eighth

(3)  > one thirteenth

TWO TRUTHS, ONE LIE

Which of the three statements below is a lie? Explain how you made your choice.

🐢 = $\dfrac{1}{3}$ 🦄 = $\dfrac{1}{5}$ 🐣 = $\dfrac{1}{8}$

(1) 🐣 + 🐣 = $\dfrac{1}{4}$

(2) 🐢 + 🐢 + 🐢 + 🐢 = $\dfrac{3}{4}$

(3) 🦄 + 🦄 + 🦄 + 🦄 + 🦄 = 1

TWO TRUTHS, ONE LIE

Which of the three statements below is a lie? Explain how you made your choice.

🍰 = $\frac{1}{5}$ ☕ = $\frac{2}{5}$ 🥐 = $\frac{4}{5}$

(1) 🥐 + 🍰 = 1 whole

(2) ☕ + ☕ = 🥐

(3) 🍰 + $\frac{4}{10}$ = ☕

TWO TRUTHS, ONE LIE

Which of the three statements below is a lie? Explain how you made your choice.

♥ = $\dfrac{9}{5}$ 🤖 = $\dfrac{7}{9}$ 💎 = $\dfrac{1}{5}$

(1)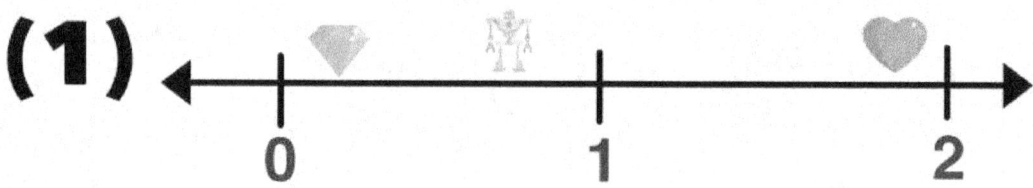

(2)

(3)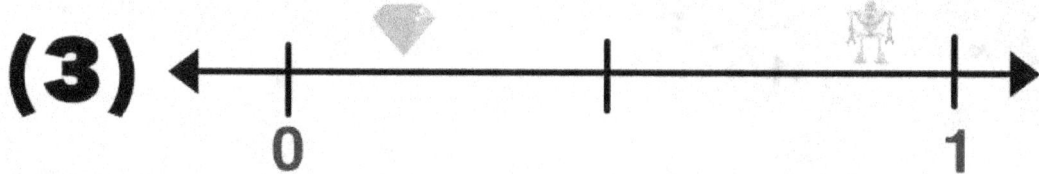

TWO TRUTHS, ONE LIE

Which of the three statements below is a lie? Explain how you made your choice.

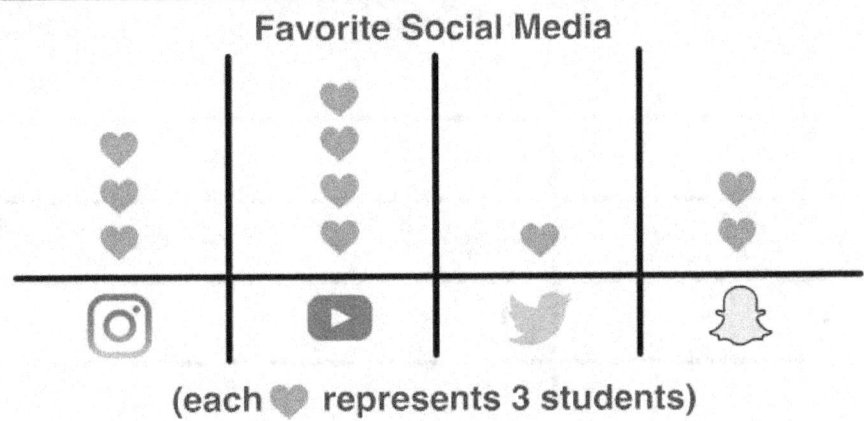

Favorite Social Media

(each ♥ represents 3 students)

(1) ▶ was chosen by the most students.

(2) 3 more students chose 📷 over 👻

(3) Only one student chose 🐦

TWO TRUTHS, ONE LIE

Which of the three statements below is a lie? Explain how you made your choice.

Favorite Pets

Pet	Number of Students
🐱	|||| |||
🐶	||||	
🐰	||
🐟	||||

(1) Two more students chose cats over dogs as their favorite pet.

(2) The chart shows a total of 21 students.

(3) Twice as many students chose fish over rabbits.

TWO TRUTHS, ONE LIE

Which of the three statements below is a lie? Explain how you made your choice.

Favorite Television Channels

Channel	Number of Students
Disney	🔲🔲🔲🔲
Nickelodeon	🔲🔲🔲🔲🔲🔲
Discovery	🔲🔲🔲🔲
Animal Planet	🔲🔲

Each 🔲 represents 4 students.

(1) Disney and Nickelodeon were chosen by the same number of students.

(2) 6 students chose Animal Planet.

(3) 12 more students chose Nickelodeon over Discovery.

4TH GRADE

TWO TRUTHS, ONE LIE

Which of the three statements below is a lie? Explain how you made your choice.

(1) 152,000

15 ten thousands + 2 thousands

(2) 27,305

27 thousands + 3 hundreds + 5 ones

(3) 2,402,000

2 millions + 4 hundred thousands + 2 ones

TWO TRUTHS, ONE LIE

Which of the three statements below is a lie? Explain how you made your choice.

(1) 27,085

20,000 + 7,000 + 80 + 5

(2) 270,850

200,000 + 70,000 + 800 + 5

(3) 909,752

900,000 + 9,000 + 700 + 50 + 2

TWO TRUTHS, ONE LIE

Which of the three statements below is a lie? Explain how you made your choice.

(1) 242,017 < 97,989

(2) 8,200 > 8,199

(3) 258,154 > 243,878

TWO TRUTHS, ONE LIE

Which of the three statements below is a lie? Explain how you made your choice.

(1) 8 hundreds
5 ten thousands   >   3 ten thousands
9 ones   9 hundreds
                      9 ones

(2) 5 ten thousands
5 thousands
5 hundreds   =   55,540
4 tens

(3) 2 hundreds
5 ten thousands   >   5 ten thousands
7 ones   2 ones
                      3 hundreds

TWO TRUTHS, ONE LIE

Which of the three statements below is a lie? Explain how you made your choice.

Mountain	Elevation
Makalu	27,825 feet
Mount Everest	29,029 feet
Mount Washington	6,289 feet
Cho Oyu	26,864 feet
Manaslu	26,759 feet
Mount Agung	9,944 feet
Mount Elbert	14,439 feet
K2	28,251 feet

(1)
Mount Elbert is shorter than K2, but taller than Mount Agung.

(2)
Cho Oyu us taller than Manaslu, but shorter than K2.

(3)
Mansalu is shorter than K2, but taller than Cho Oyu.

TWO TRUTHS, ONE LIE

Which of the three statements below is a lie? Explain how you made your choice.

WEIGHT

1 kilogram   1,000 grams   500 kilograms

(1)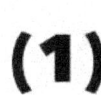

(2)

(3)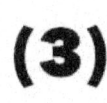

TWO TRUTHS, ONE LIE

Which of the three statements below is a lie? Explain how you made your choice.

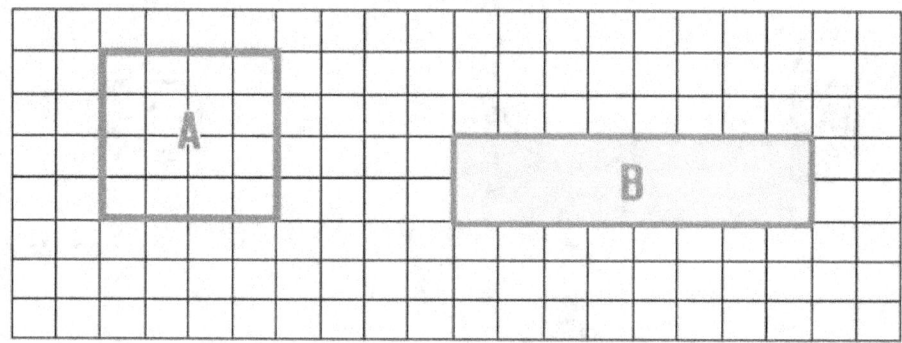

(1) Figure A and figure B have the same area.

(2) Figure A and figure B have the same perimeter.

(3) The area of figure A is equal to its perimeter.

TWO TRUTHS, ONE LIE

Which of the three statements below is a lie? Explain how you made your choice.

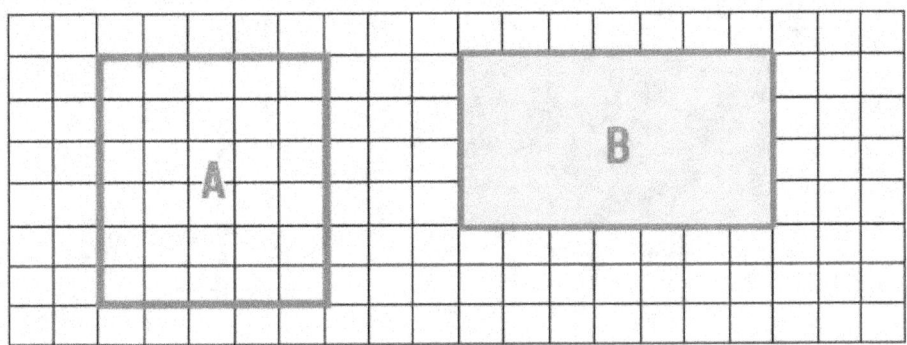

(1) The perimeter of figure B is greater than the area of figure A.

(2) The area of figure A is greater than the perimeter of figure B.

(3) Figure A and figure B have the same perimeter.

TWO TRUTHS, ONE LIE

Which of the three statements below is a lie? Explain how you made your choice.

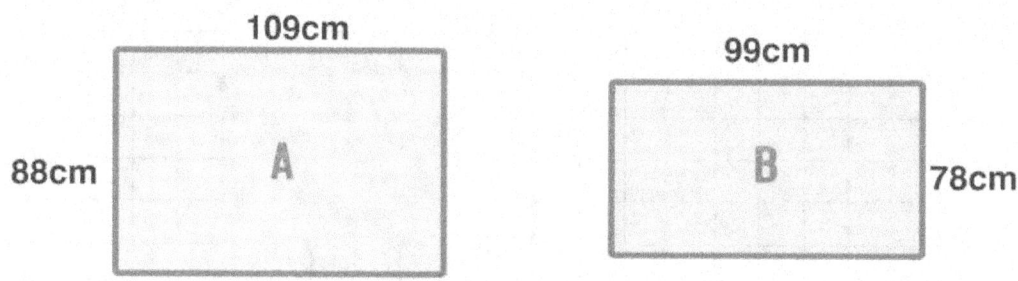

(1) Figure A has a perimeter of 394 cm.

(2) The perimeter of figure B is 40 cm larger than the perimeter of figure A.

(3) The area of figure B is 7,722 square centimeters.

TWO TRUTHS, ONE LIE

Which of the three statements below is a lie? Explain how you made your choice.

FIGURE A
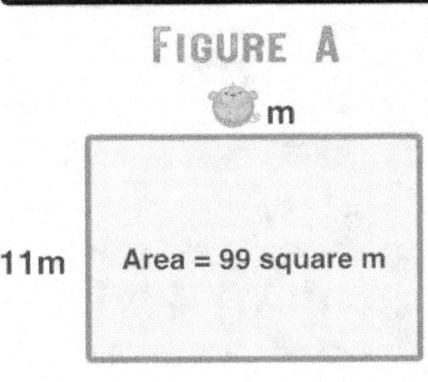

11m, 🐵 m, Area = 99 square m

FIGURE B
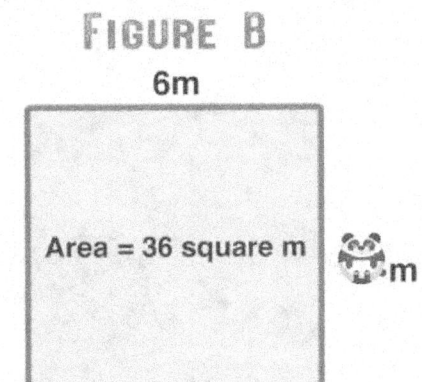

6m, 🐼 m, Area = 36 square m

(1) 🐼 + 3 = 🐵

(2) 🐼 > 🐵

(3) 🐵 > 🐼

TWO TRUTHS, ONE LIE

Which of the three statements below is a lie? Explain how you made your choice.

$$7 \times \text{🐄} = 700$$

$$\text{🐄} \times \text{🐂} = 1{,}000$$

(1) $\text{🐄} = 100$

(2) $\text{🐄} + \text{🐂} = 1{,}010$

(3) $10 = \text{🐂}$

TWO TRUTHS, ONE LIE

Which of the three statements below is a lie? Explain how you made your choice.

$$14 \times 🐷 = 1{,}400$$

$$1{,}400 = 14 \times 🦄 \times 🐥$$

$$🦄 = 🐥$$

(1) 🐷 = 100

(2) 🦄 × 🐥 > 🐷

(3) 🦄 + 🐥 = 20

TWO TRUTHS, ONE LIE

Which of the three statements below is a lie? Explain how you made your choice.

$20 \times 🐻 = 1{,}000$

$🐻 \times 30 = 1{,}500$

$🐻 + 🐻 = 100$

(1) 🐻 and 🐻 do not equal the same value.

(2) 🐻 = 50

(3) $2{,}500 = 🐻 \times 🐻$

TWO TRUTHS, ONE LIE

Which of the three statements below is a lie? Explain how you made your choice.

(1) $9 \times 7{,}000 = 63{,}000$

(2) $76{,}000 = 9{,}000 \times 9$

(3) $56{,}000 = 8 \times 7{,}000$

TWO TRUTHS, ONE LIE

Which of the three statements below is a lie? Explain how you made your choice.

(1) Hundreds: 4, Tens: 7, Ones: 9 → 479

(2) Hundreds: 0, Tens: 11, Ones: 11 → 111

(3) Hundreds: 4, Tens: 12, Ones: 5 → 525

TWO TRUTHS, ONE LIE

Which of the three statements below is a lie? Explain how you made your choice.

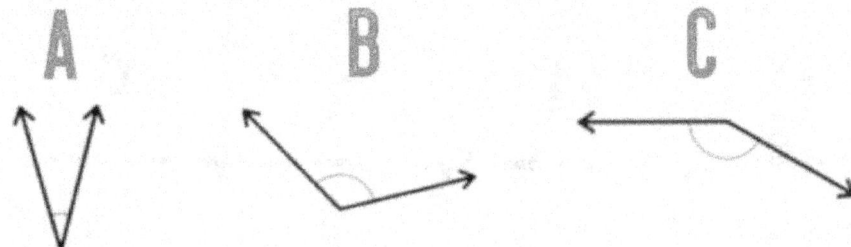

(1) A and B are acute angles.

(2) The measure of C is greather than 90 degrees.

(3) C and B are obtuse angles.

TWO TRUTHS, ONE LIE

Which of the three statements below is a lie? Explain how you made your choice.

(1) $\dfrac{11}{7} = \dfrac{1}{7} + \dfrac{4}{7} + \dfrac{6}{7}$

(2) $2 = \dfrac{1}{4} + \dfrac{3}{4} + \dfrac{4}{2}$

(3) $\dfrac{1}{5} + \dfrac{1}{5} + \dfrac{2}{5} = \dfrac{4}{5}$

TWO TRUTHS, ONE LIE

Which of the three statements below is a lie? Explain how you made your choice.

(1) $\dfrac{13}{9} = \dfrac{7}{9} + \dfrac{2}{9} + \dfrac{4}{9}$

(2) $3\dfrac{1}{3} = \dfrac{3}{3} + \dfrac{1}{3}$

(3) $3\dfrac{5}{8} = 3 + \dfrac{1}{8} + \dfrac{3}{8} + \dfrac{1}{8}$

TWO TRUTHS, ONE LIE

Which of the three statements below is a lie? Explain how you made your choice.

(1) $\dfrac{12}{8} = \dfrac{3}{8} + \dfrac{7}{8} + \dfrac{4}{8}$

(2) $2\dfrac{2}{7} = \dfrac{2}{7} + 2$

(3) $\dfrac{12}{11} = 1 + \dfrac{1}{11}$

TWO TRUTHS, ONE LIE

Which of the three statements below is a lie? Explain how you made your choice.

(1) $\dfrac{1}{2} = \dfrac{2}{4}$

(2) $\dfrac{1}{3} = \dfrac{2}{6}$

(3) $\dfrac{2}{3} = \dfrac{4}{7}$

TWO TRUTHS, ONE LIE

Which of the three statements below is a lie? Explain how you made your choice.

(1) $\dfrac{1}{2} > \dfrac{3}{4}$

(2) $\dfrac{8}{12} = \dfrac{4}{6}$

(3) $\dfrac{3}{7} > \dfrac{4}{14}$

TWO TRUTHS, ONE LIE

Which of the three statements below is a lie? Explain how you made your choice.

🫘 = 3   🧸 = 2   🍉 = 10

(1) $\dfrac{1}{4} > \dfrac{1}{🧸}$

(2) $\dfrac{8}{🍉} > \dfrac{🫘}{5}$

(3) $\dfrac{1}{🧸} < \dfrac{🫘}{5}$

TWO TRUTHS, ONE LIE

Which of the three statements below is a lie? Explain how you made your choice.

🏰 = 3   🚐 = 5   🏐 = 6

(1) $\dfrac{1}{2} < \dfrac{🏰}{🚐}$

(2) $\dfrac{🏰}{1} = \dfrac{2}{🏐}$

(3) $\dfrac{🚐}{12} < \dfrac{🚐}{8}$

TWO TRUTHS, ONE LIE

Which of the three statements below is a lie? Explain how you made your choice.

🧑‍🦲 = 6   👻 = 8   🎮 = 100

(1) $\dfrac{\text{🧑‍🦲}}{9} < \dfrac{2}{3}$

(2) $\dfrac{4}{\text{🧑‍🦲}} = \dfrac{\text{👻}}{12}$

(3) $\dfrac{49}{\text{🎮}} < \dfrac{5}{10}$

TWO TRUTHS, ONE LIE

Which of the three statements below is a lie? Explain how you made your choice.

🍔 = 7   🧁 = 8   🌮 = 12

(1) $\dfrac{3}{🧁} < \dfrac{🍔}{🌮}$

(2) $\dfrac{5}{🌮} > \dfrac{🍔}{🧁}$

(3) $\dfrac{🧁}{6} > \dfrac{11}{🌮}$

TWO TRUTHS, ONE LIE

Which of the three statements below is a lie? Explain how you made your choice.

🎾 = 7   🏀 = 2   🎳 = 1

(1) $\dfrac{5}{3} = \dfrac{3}{3} + \dfrac{🏀}{3}$

(2) $\dfrac{🎾}{4} = \dfrac{3}{4} + \dfrac{4}{4}$

(3) $\dfrac{🎾}{5} = \dfrac{5}{5} + \dfrac{🎳}{5}$

TWO TRUTHS, ONE LIE

Which of the three statements below is a lie? Explain how you made your choice.

🍉 = 10

(1) $180 \div 🍉 = 18$

(2) $6{,}300 \div 🍉 = 63$

(3) $10{,}500 \div 🍉 = 1{,}050$

TWO TRUTHS, ONE LIE

Which of the three statements below is a lie? Explain how you made your choice.

🍔 = 10

(1) $0.7 = \dfrac{7}{🍔}$

(2) $2.6 = 2 + \dfrac{6}{🍔}$

(3) $3.5 = 3 + \dfrac{1}{5}$

TWO TRUTHS, ONE LIE

Which of the three statements below is a lie? Explain how you made your choice.

🫐 = 7   🫐🫐 = 12   🍎 = 25

(1) $0.1 \times 🫐 = 0.5$

(2) $0.1 \times 🫐🫐 = 1.2$

(3) $0.1 \times 🍎 = 2.5$

TWO TRUTHS, ONE LIE

Which of the three statements below is a lie? Explain how you made your choice.

(1) $2\frac{7}{10} = 7.2$

(2) $3.5 = 3\frac{5}{10}$

(3) $\frac{40}{10} = 4$

TWO TRUTHS, ONE LIE

Which of the three statements below is a lie? Explain how you made your choice.

(1) 2 km = 2,000 m

(2) 1,000 cm = 1 m

(3) 84 in = 7 ft

TWO TRUTHS, ONE LIE

Which of the three statements below is a lie? Explain how you made your choice.

(1) 2 quarts > 4 pints

(2) 1 gallon = 4 quarts

(3) 1 quart = 4 cups

TWO TRUTHS, ONE LIE

Which of the three statements below is a lie? Explain how you made your choice.

24oz   16oz   8oz

(1) 🥣 + 🥣 = 2 lb

(2) 🥞 − 🧁 = 1 lb

(3) 🧁 + 🧁 + 🥞 = 3 lb

TWO TRUTHS, ONE LIE

Which of the three statements below is a lie? Explain how you made your choice.

(1) There are 16 cups in one gallon.

(2) There are 64 quarts in 16 gallons.

(3) There are 38 ounces in 2 pounds.

TWO TRUTHS, ONE LIE

Which of the three statements below is a lie? Explain how you made your choice.

(1) $3 \text{ inches} = \frac{1}{3} \text{ feet}$

(2) $6 \text{ inches} = \frac{1}{2} \text{ feet}$

(3) $30 \text{ inches} = 2\frac{1}{2} \text{ feet}$

5TH GRADE

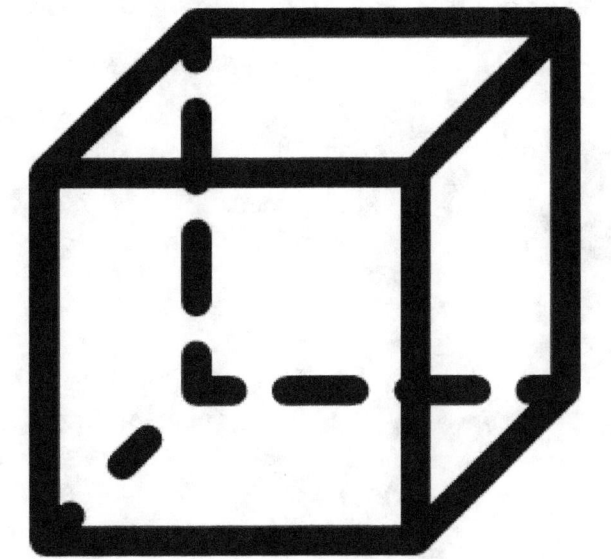

TWO TRUTHS, ONE LIE

Which of the three statements below is a lie? Explain how you made your choice.

(1) $5 \times 10^2 = 50$

(2) $9 \times 10^3 = 9000$

(3) $24 \times 10^4 = 240{,}000$

TWO TRUTHS, ONE LIE

Which of the three statements below is a lie? Explain how you made your choice.

(1) $2.16 \times 10^4 = 2{,}160$

(2) $3 \times 10^3 = 3{,}000$

(3) $3.21 \times 10^4 = 32{,}100$

TWO TRUTHS, ONE LIE

Which of the three statements below is a lie? Explain how you made your choice.

(1) $2\text{m} \times 10^3 = 2{,}000\text{m}$

(2) $2\text{m} \times 10^3 = 2{,}000\text{mm}$

(3) $2\text{m} = 2{,}000\text{mm}$

TWO TRUTHS, ONE LIE

Which of the three statements below is a lie? Explain how you made your choice.

(1) $78.3 \times 1{,}000 = 78{,}300$

(2) $78.3 \times 100 = 78{,}300$

(3) $78.3 \times 10 = 783.0$

TWO TRUTHS, ONE LIE

Which of the three statements below is a lie? Explain how you made your choice.

(1) $0.003 \times 100 = 0.3$

(2) $0.03 \times 100 = 0.3$

(3) $0.3 \times 10 = 3.0$

TWO TRUTHS, ONE LIE

Which of the three statements below is a lie? Explain how you made your choice.

(1) $16.45 < 16.4$

(2) $0.83 = \dfrac{83}{100}$

(3) $\dfrac{205}{1000} = 0.205$

TWO TRUTHS, ONE LIE

Which of the three statements below is a lie? Explain how you made your choice.

Least to Greatest →

(1) 6.038   6.3   7.401   7.41

(2) 500.01   500.26   500.401   500.049

(3) 31.25   31.26   31.30   31.9

TWO TRUTHS, ONE LIE

Which of the three statements below is a lie? Explain how you made your choice.

(1) $0.3 + 0.82 = 0.85$

(2) $7.8 + 3.22 = 11.02$

(3) $2.49 + 0.01 = 2.50$

TWO TRUTHS, ONE LIE

Which of the three statements below is a lie? Explain how you made your choice.

(1) $9 \times 3 = (5 \times 3) + (4 \times 3)$

(2) $7 \times 4 = (2 \times 7) + (2 \times 7)$

(3) $8 \times 8 = (8 \times 5) + (8 \times 4)$

TWO TRUTHS, ONE LIE

Which of the three statements below is a lie? Explain how you made your choice.

🎃 = 4   🧟 = 10   🎩 = 21

(1) The sum of  and , doubled, is 50.

(2) Five times the sum of  and  is 140.

(3) Triple the sum of  and  is 93.

TWO TRUTHS, ONE LIE

Which of the three statements below is a lie? Explain how you made your choice.

🦀 = 3   🐢 = 6   🐦 = 9

(1) Four times the difference of 🐦 and 🦀 is 12.

(2) The sum of 🐢 and 🦀, tripled, is 27.

(3) Twice the difference of 🐦 and 🐢 is 6.

TWO TRUTHS, ONE LIE

Which of the three statements below is a lie? Explain how you made your choice.

(1)

	40	8
20	800	🍓
2		16

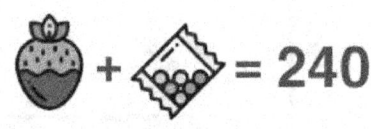 + 🍬 = 240

(2)

	50	39
40	🍩	1560
5	250	🍰

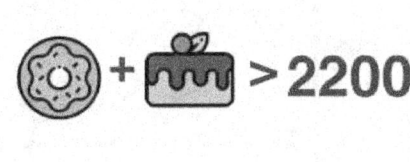 🍩 + 🍰 > 2200

(3)

	70	4
50	🍫	🍦
2	140	8

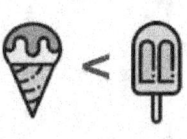 🍦 < 🍫

TWO TRUTHS, ONE LIE

Which of the three statements below is a lie? Explain how you made your choice.

(1)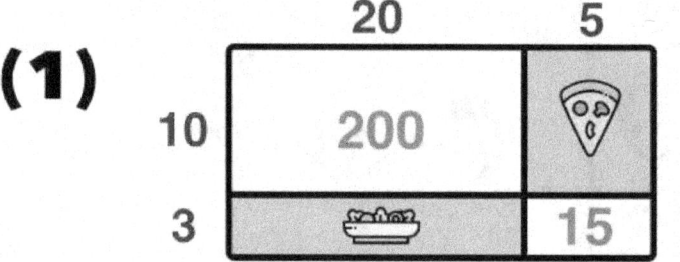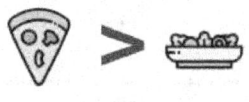

(2)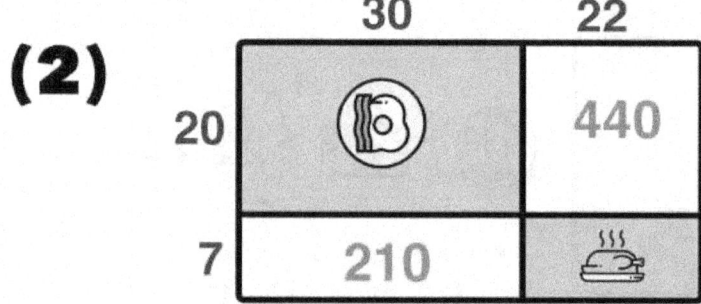

(3)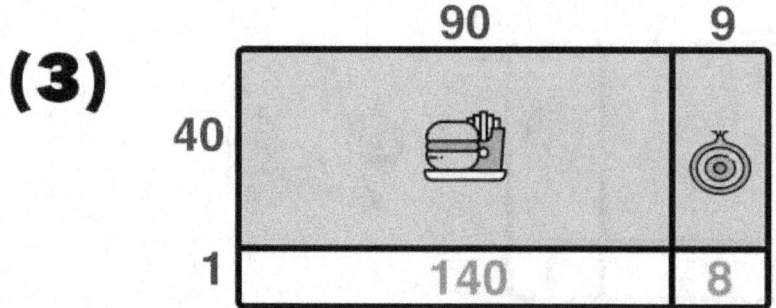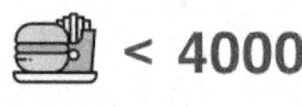

TWO TRUTHS, ONE LIE

Which of the three statements below is a lie? Explain how you made your choice.

(1)

	10	3
5	50	🐧
4	🎮	12

🐧 < 🎮

(2)

	10	7
5	🚀	35
4	40	🏐

🚀 + 🏐 = 78

(3)

	20	2
10	🧸	20
1	🧊	2

🧸 + 🧊 > 225

TWO TRUTHS, ONE LIE

Which of the three statements below is a lie? Explain how you made your choice.

(1) There are 28 days in four weeks.

(2) There are 4,000 grams in 4 kilograms.

(3) There are 120 hours in 7 days.

TWO TRUTHS, ONE LIE

Which of the three statements below is a lie? Explain how you made your choice.

(1) $100 \times 8 < 25 \times (4 \times 9)$

(2) $36 \times 12 = 40$ twelves $- 3$ twelves

(3) $24 \times 36 = 18$ twenty-fours $+ 18$ twenty-fours

TWO TRUTHS, ONE LIE

Which of the three statements below is a lie? Explain how you made your choice.

(1) 40 times the sum of 35 and 91

$$40 \times 35 + 91$$

(2) 25 times the difference of 1,200 and 300

$$25 \times (1{,}200 - 300)$$

(3) The sum of 3 tens and 16 tens

$$(3 \times 10) + (16 \times 10)$$

TWO TRUTHS, ONE LIE

Which of the three statements below is a lie? Explain how you made your choice.

ESTIMATION

(1) $4{,}738 \div 21$
$\approx 5{,}000 \div 20 = 250$

(2) $1{,}463 \div 53$
$\approx 1{,}500 \div 50 = 30$

(3) $7{,}901 \div 75$
$\approx 7{,}000 \div 100 = 70$

TWO TRUTHS, ONE LIE

Which of the three statements below is a lie? Explain how you made your choice.

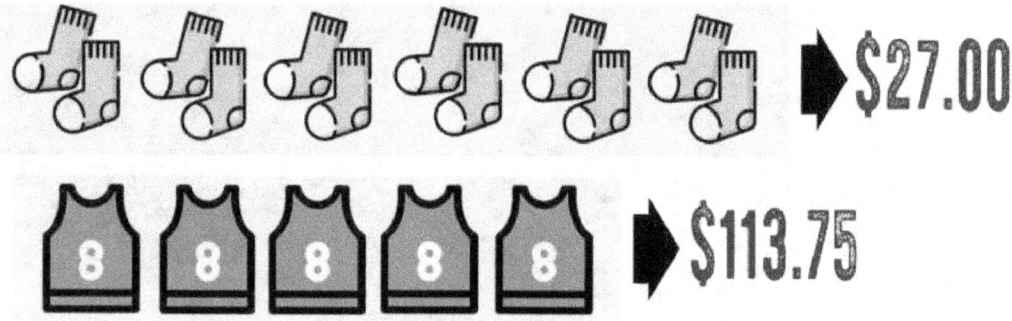

(1) Five pairs of socks costs $22.50

(2) Two jerseys costs $45.50

(3) One jersey is more expensive than 6 pairs of socks.

TWO TRUTHS, ONE LIE

Which of the three statements below is a lie? Explain how you made your choice.

EQUIVALENT FRACTIONS

(1) $\dfrac{1}{3}$ $\dfrac{10}{30}$ $\dfrac{7}{21}$

(2) $\dfrac{18}{24}$ $\dfrac{2}{3}$ $\dfrac{9}{12}$

(3) $\dfrac{55}{66}$ $\dfrac{5}{6}$ $\dfrac{35}{42}$

TWO TRUTHS, ONE LIE

Which of the three statements below is a lie? Explain how you made your choice.

EQUIVALENT FRACTIONS

(1) $\dfrac{45}{81}$ $\dfrac{9}{16}$ $\dfrac{5}{9}$

(2) $\dfrac{36}{81}$ $\dfrac{4}{9}$ $\dfrac{12}{27}$

(3) $\dfrac{9}{12}$ $\dfrac{21}{28}$ $\dfrac{75}{100}$

TWO TRUTHS, ONE LIE

Which of the three statements below is a lie? Explain how you made your choice.

EQUIVALENT FRACTIONS

(1) $\dfrac{11}{12}$ $\dfrac{22}{24}$ $\dfrac{33}{36}$

(2) $\dfrac{4}{5}$ $\dfrac{36}{45}$ $\dfrac{44}{54}$

(3) $\dfrac{27}{45}$ $\dfrac{15}{25}$ $\dfrac{3}{5}$

TWO TRUTHS, ONE LIE

Which of the three statements below is a lie? Explain how you made your choice.

(1) $2 - 1\frac{3}{8} = \frac{3}{8}$

(2) $3 - 1\frac{1}{4} = 1\frac{3}{4}$

(3) $7\frac{3}{4} - 2\frac{1}{4} = 5\frac{1}{2}$

TWO TRUTHS, ONE LIE

Which of the three statements below is a lie? Explain how you made your choice.

(1) $6 + 3\dfrac{1}{2} = 9\dfrac{1}{2}$

(2) $5 - 2\dfrac{2}{3} = 2\dfrac{1}{3}$

(3) $7\dfrac{3}{4} - 4\dfrac{3}{4} = 3\dfrac{1}{4}$

TWO TRUTHS, ONE LIE

> Which of the three statements below is a lie?
> Explain how you made your choice.

(1) $6 + 3\frac{5}{6} = 9\frac{5}{6}$

(2) $6 - 2\frac{1}{2} = 3\frac{1}{2}$

(3) $6\frac{1}{5} - 3\frac{4}{5} = 2\frac{1}{5}$

TWO TRUTHS, ONE LIE

Which of the three statements below is a lie? Explain how you made your choice.

🦔 $= 3\frac{3}{5}$ 🐇 $= 5\frac{1}{4}$ 🐢 $= 2\frac{1}{2}$

(1) Number line from 2 to 6 with 🐢 between 2 and 3, 🦔 between 3 and 4 (near 4), 🐇 between 5 and 6.

(2) 🦔 + 🐢 < 🐇

(3) $7\frac{3}{4}$ = 🐢 + 🐇

TWO TRUTHS, ONE LIE

Which of the three statements below is a lie? Explain how you made your choice.

🎯 = 6.1   🎾 = 4.5   🏐 = 5.33

(1)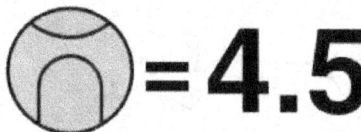

(2) 🎾 + 🏐 > 🎯 + 3

(3) 9.83 = 🎾 + 🏐

TWO TRUTHS, ONE LIE

Which of the three statements below is a lie? Explain how you made your choice.

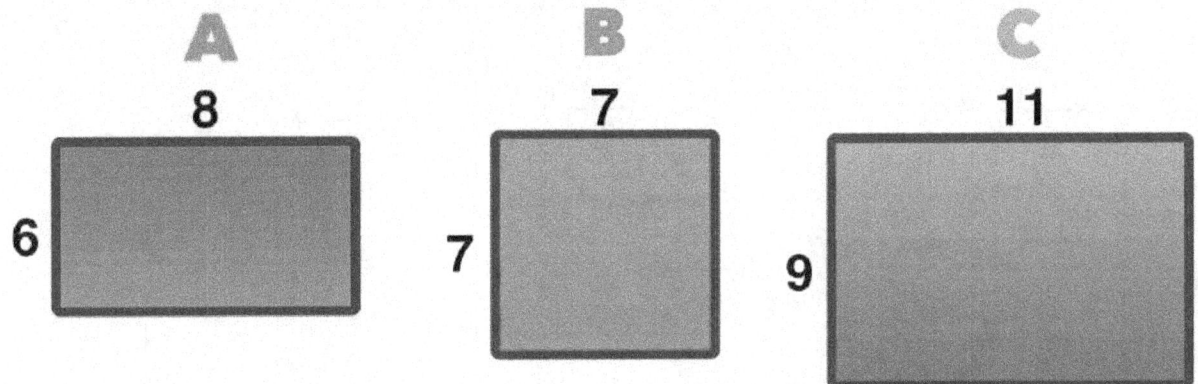

(1) C has the largest area

(2) The area of A is larger than the area of B.

(3) The area of B is 49 square units.

TWO TRUTHS, ONE LIE

Which of the three statements below is a lie? Explain how you made your choice.

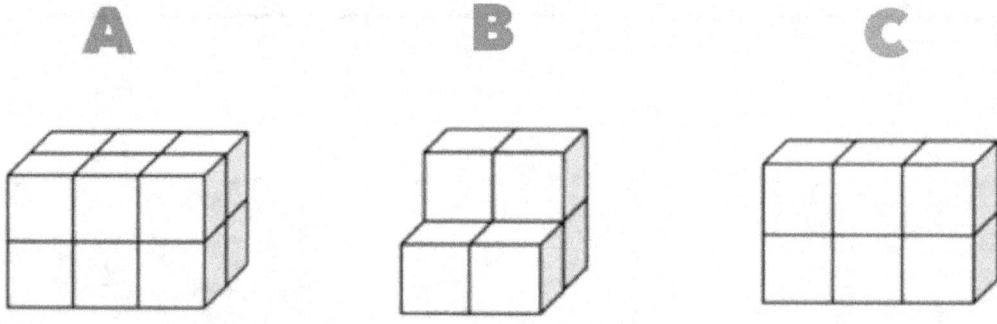

A   B   C

(1) A has a volume of 16 square units.

(2) The volume of A is double the volume of C

(3) B and C have the same volume.

TWO TRUTHS, ONE LIE

Which of the three statements below is a lie? Explain how you made your choice.

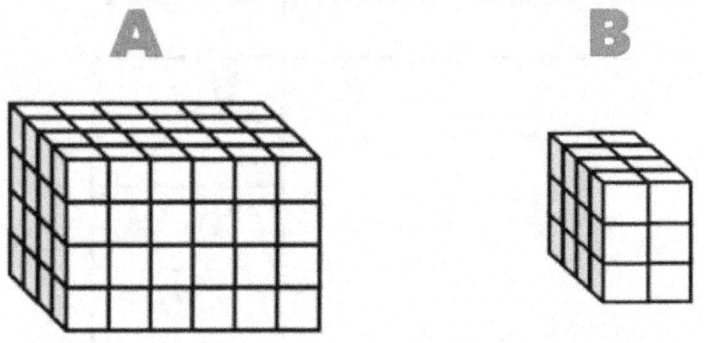

(1) The volume of B is one-third the volume of A

(2) A has a volume of 84 square units.

(3) A has a volume of 96 square units.

TWO TRUTHS, ONE LIE

Which of the three statements below is a lie? Explain how you made your choice.

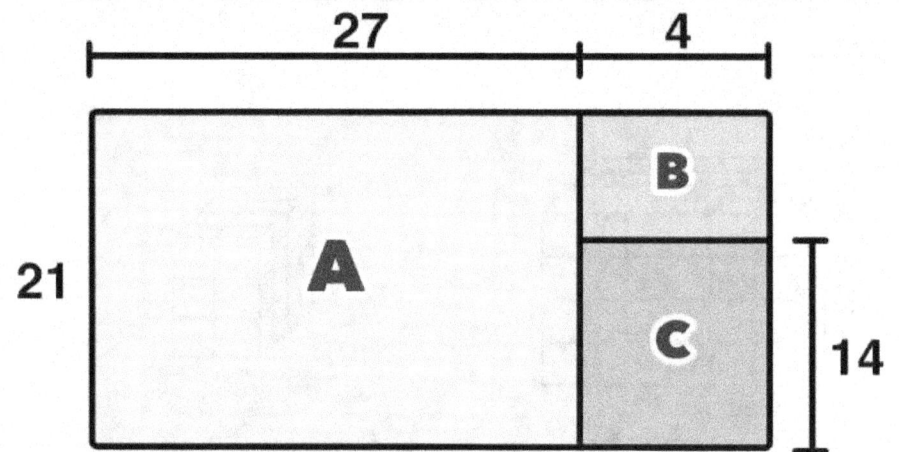

(1) The area of **A** is 567 square units

(2) The area of **C** is 56 square units

(3) The area of **B** is 16 square units

TWO TRUTHS, ONE LIE

Which of the three statements below is a lie? Explain how you made your choice.

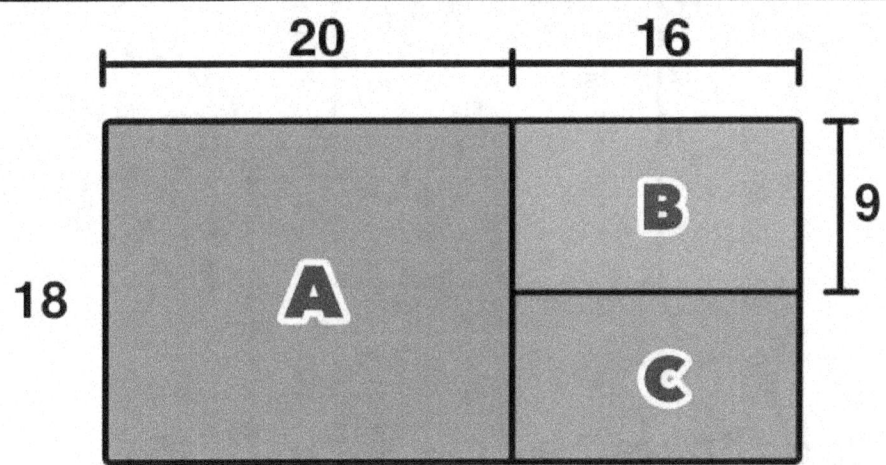

(1) The area of A is 360 square units

(2) The area of C is 144 square units

(3) The area of B is greater than the volume of C

TWO TRUTHS, ONE LIE

Which of the three statements below is a lie? Explain how you made your choice.

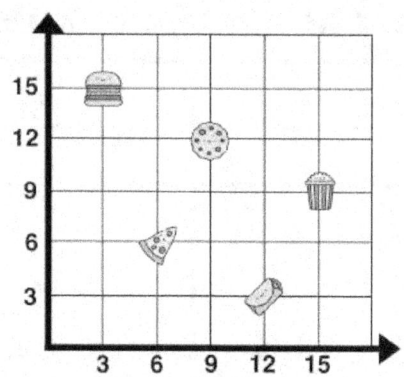

(1) The coordinate pair of 🍿 is (9,15)

(2) The coordinate pair of 🍔 is (3,15)

(3) The coordinate pair of 🍕 is (6,6)

TWO TRUTHS, ONE LIE

Which of the three statements below is a lie? Explain how you made your choice.

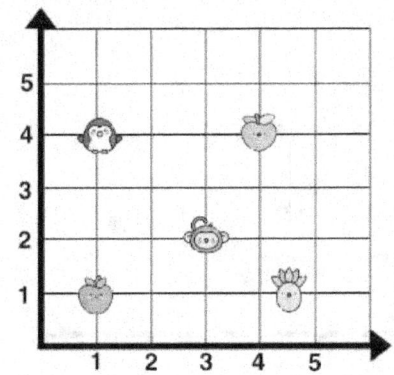

(1) The coordinate pair of 🐵 is (3,2)

(2) The coordinate pair of 🌱 is ($4\frac{1}{2}$, 1)

(3) The coordinate pair of 🐧 is (4,1)

TWO TRUTHS, ONE LIE

Which of the three statements below is a lie? Explain how you made your choice.

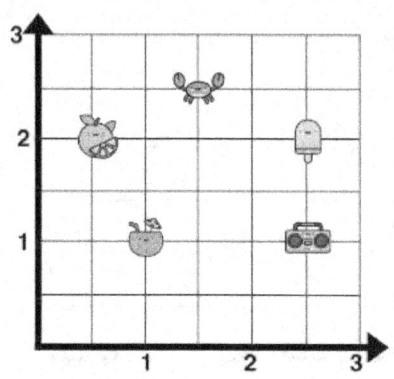

(1) The coordinate pair of 🥥 is (1,1)

(2) The coordinate pair of 🍦 is $(2\frac{1}{2}, 2\frac{1}{2})$

(3) The coordinate pair of 🦀 is $(1\frac{1}{2}, 2\frac{1}{2})$

ANSWER KEY

3rd Grade
01) 1
02) 3
03) 2
04) 3
05) 2
06) 1
07) 1
08) 1
09) 2
10) 2
11) 2
12) 3
13) 2
14) 2
15) 2
16) 2
17) 1
18) 3
19) 3
20) 2
21) 3
22) 1
23) 1
24) 2
25) 3
26) 2
27) 3
28) 3
29) 2
30) 3
31) 2
32) 3
33) 2
34) 1

4th Grade
01) 3
02) 2
03) 1
04) 3
05) 3
06) 3
07) 2
08) 1
09) 2
10) 2
11) 2
12) 2
13) 1
14) 2
15) 2
16) 1
17) 2
18) 2
19) 1
20) 3
21) 1
22) 1
23) 2
24) 1
25) 2
26) 3
27) 2
28) 3
29) 1
30) 1
31) 2
32) 1
33) 3
34) 3
35) 1

5th Grade
01) 1
02) 1
03) 2
04) 2
05) 2
06) 1
07) 2
08) 1
09) 3
10) 2
11) 1
12) 2
13) 1
14) 3
15) 3
16) 2
17) 1
18) 3
19) 3
20) 2
21) 1
22) 2
23) 1
24) 3
25) 3
26) 2
27) 1
28) 2
29) 1
30) 2
31) 3
32) 3
33) 1
34) 3
35) 2

About Mashup Math

MashUp Math is your go-to source for resources, activities, and ideas for making math education fun and exciting for your kids every day! Whether you're looking for lesson plan ideas, free resources, puzzles, and worksheets, video lessons, research and insights, or to connect with other educators, you will always find something cool, fun, and useful at **www.mashupmath.com**!

Looking for free daily resources, math puzzles, fun ideas, and more?
Follow us on social media!

WAIT!
Do You Want Free Math Resources In Your Inbox Every Day?

Visit www.mashupmath.com to subscribe to our mailing list to get free math resources, puzzles and activities, special promotions, and more in your inbox every week!

www.ingramcontent.com/pod-product-compliance
Lightning Source LLC
Chambersburg PA
CBHW080500220526
45465CB00006B/2331